Visit
https://volersystems.com/category/design-tips
for more information on project management
of design and sample documents. We can help
you complete your electronic design or firmware
project on time and on budget using the project
management techniques in the book.

Highly Successful Engineering Design Projects

Keys to Staying on Budget, on Time, Every Time

Walt Maclay

THiNK_aha_®

An Actionable Engineering Design Journal

E-mail: info@thinkaha.com
20660 Stevens Creek Blvd., Suite 210
Cupertino, CA 95014

Published by THiNKaha®
20660 Stevens Creek Blvd., Suite 210, Cupertino, CA 95014
http://thinkaha.com
E-mail: info@thinkaha.com

First Printing: July 2019
Hardcover ISBN: 978-1-61699-323-8   1-61699-323-5
Paperback ISBN: 978-1-61699-322-1   1-61699-322-7
eBook ISBN: 978-1-61699-321-4   1-61699-321-9
Place of Publication: Silicon Valley, California, USA
Paperback Library of Congress Number: 2019907893

Trademarks

Warning and Disclaimer

Acknowledgement

John Dring, VP of engineering at Voler Systems, contributed much of his great knowledge and experience. He consistently completes projects on time and on budget, and the products are easy to manufacture. John is well qualified to contribute!

Kimberly Wiefling, founder and president of Wiefling Consulting, edited and contributed a great deal. She is a noted project manager who advises design teams around the world.

How to Read a THiNKaha® Book

A Note from the Publisher

The AHAthat/THiNKaha series is the CliffsNotes of the 21st century. These books are contextual in nature. Although the actual words won't change, their meaning will every time you read one as your context will change. Be ready, you will experience your own AHA moments as you read the AHA messages™ in this book. They are designed to be stand-alone actionable messages that will help you think about a project you're working on, an event, a sales deal, a personal issue, etc. differently. As you read this book, please think about the following:

1. It should only take 15–20 minutes to read this book the first time out. When you're reading, write in the underlined area one to three action items that resonate with you.
2. Mark your calendar to re-read this book again in 30 days.
3. Repeat step #1 and mark one to three more AHA messages that resonate. They will most likely be different than the first time. BTW: this is also a great time to reflect on the AHA messages that resonated with you during your last reading.

After reading a THiNKaha book, marking your AHA messages, re-reading it, and marking more AHA messages, you'll begin to see how these books contextually apply to you. AHAthat/THiNKaha books advocate for continuous, lifelong learning. They will help you transform your AHAs into actionable items with tangible results until you no longer have to say AHA to these moments—they'll become part of your daily practice as you continue to grow and learn.

Mitchell Levy, The AHA Guy at AHAthat
publisher@thinkaha.com

THiNKaha®

Contents

Highly successful #EngineeringDesign projects make the company successful. #SuccessfulProjects

Walt Maclay
http://aha.pub/EngineeringDesignProjects

Share the AHA messages from this book socially by going to
http://aha.pub/EngineeringDesignProjects

Section I

Characteristics of Highly Successful Engineering Design Projects

How do you know if you have a successful engineering design project or if you have room to improve? Learn more about what a successful project looks like.

1

Everyone can learn to have highly successful #EngineeringDesign projects. #SuccessfulProjects

_____

_____

_____

2

Highly successful #EngineeringDesign projects are completed on time. #SuccessfulProjects

_____

_____

_____

3

Highly successful #EngineeringDesign projects are completed on budget. #SuccessfulProjects

_____

_____

_____

4

Highly successful #EngineeringDesign projects result in products that are easy to manufacture. #SuccessfulProjects

_____

_____

_____

5

Highly successful #EngineeringDesign projects are adaptable to change. #SuccessfulProjects

_____

_____

_____

6

Scalable #EngineeringDesign projects don't require re-design to adapt and grow. #SuccessfulProjects

_____

_____

_____

7

Highly successful #EngineeringDesign projects result in high-quality products. #SuccessfulProjects

_____

_____

_____

8

Highly successful #EngineeringDesign projects make team members happy. #SuccessfulProjects

_____

_____

_____

9

Highly successful #EngineeringDesign projects make management happy. #SuccessfulProjects

_____

_____

_____

10

Highly successful #EngineeringDesign projects make the project leader happy and successful. #SuccessfulProjects

_____

_____

_____

11

Highly successful #EngineeringDesign projects become the models for other projects. #SuccessfulProjects

_____

_____

_____

12

Highly successful #EngineeringDesign projects make the company successful. #SuccessfulProjects

_____

_____

_____

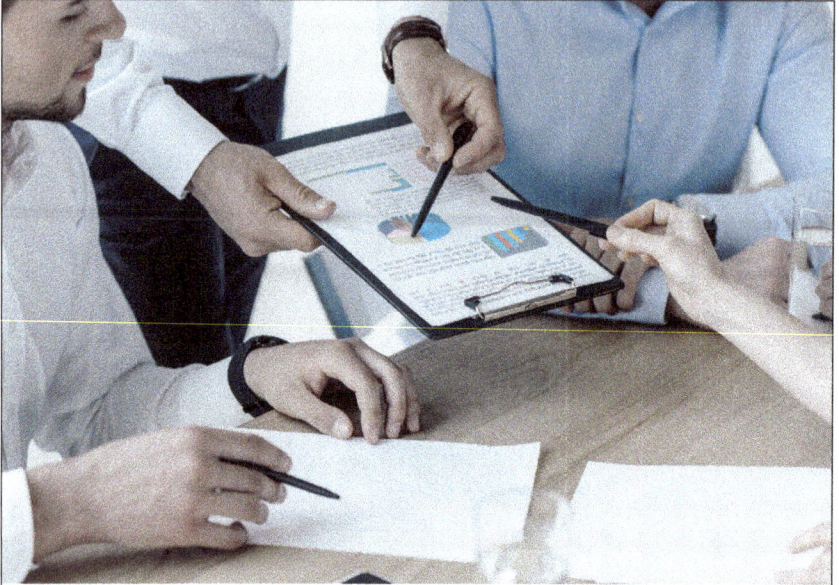

Ensure that #EngineeringDesign requirements are complete, and review them several times with your team. #SuccessfulProjects

Walt Maclay
http://aha.pub/EngineeringDesignProjects

Share the AHA messages from this book socially by going to
http://aha.pub/EngineeringDesignProjects

Section II

Critical Steps to Take before Starting a Project

There are certain steps to take at the start of your engineering design project. Without them, it is difficult to have a successful project. But by doing these things, you are already halfway there!

13

Write detailed #EngineeringDesign requirements. #SuccessfulProjects have detailed requirements that include the customer's POV (inputs) and the engineering department's POV (outputs).

_____

_____

_____

14

Make sure that #EngineeringDesign requirements are testable. To have #SuccessfulProjects, it must be clear when requirements are or are not met.

_____

_____

_____

15

Get agreement from all stakeholders on the #EngineeringDesign requirements before proceeding. #SuccessfulProjects

_____

_____

_____

16

Ensure that #EngineeringDesign requirements are complete, and review them several times with your team. #SuccessfulProjects

_____

_____

_____

17

#EngineeringDesign requirements may change as customers, competitors, and technology change. #SuccessfulProjects

_____

_____

_____

18

Describe the appearance of mechanical, electrical, software, and firmware features in the #EngineeringDesign requirements. #SuccessfulProjects

_____

_____

_____

19

Specify environmental conditions, such as temperature, humidity, altitude, shock, and vibration, in the #EngineeringDesign requirements. #SuccessfulProjects

---

20

Specify serviceability and support in the #EngineeringDesign requirements. #SuccessfulProjects

---

21

Specify cleaning and sterilization in the #EngineeringDesign requirements. #SuccessfulProjects

_____

_____

_____

22

Specify compatibility in the #EngineeringDesign requirements, such as working with other devices and software or backward compatibility with older products. #SuccessfulProjects

_____

_____

_____

23

Specify the cost target in the #EngineeringDesign requirements. #SuccessfulProjects

---

24

Writing software requirements for #EngineeringDesign projects is hard. If you have not done it before, seek advice. #SuccessfulProjects

---

25

Include pictures of each screen
in the software requirements of
your #EngineeringDesign project.
#SuccessfulProjects

_____

_____

_____

26

Describe what happens when you use each tool on your screen (e.g., buttons, text-entry boxes, drop-down lists, etc.) in the #EngineeringDesign software requirements. #SuccessfulProjects

_____

_____

_____

27

Describe how to move from one screen
to the next in the #EngineeringDesign
software requirements. #SuccessfulProjects

_____

_____

_____

28

Describe any calculations, data processing,
file processing, data transmission, or
alarms that happen and what triggers them
to happen in the #EngineeringDesign
software requirements. #SuccessfulProjects

_____

_____

_____

29

Specify the wording for each alarm in
the #EngineeringDesign requirements.
#SuccessfulProjects

_____

_____

_____

30

#EngineeringDesign requirements should be modular and traceable to higher level requirements. #SuccessfulProjects

_____

_____

_____

31

When writing requirements, understand both the user and the purchaser to have a successful #EngineeringDesign project. #SuccessfulProjects

_____

_____

_____

32

Write a detailed development schedule, with each activity covering one to two weeks, four weeks being the maximum. #EngineeringDesign #SuccessfulProjects

_____

_____

_____

33

An activity in the #EngineeringDesign schedule should be done by one person so responsibility is clear. #SuccessfulProjects

_____

_____

_____

34

Build #EngineeringDesign schedules bottom-up instead of top-down. It's okay to start at the top to provide an outline for the team, then build up the schedule from the team details. #SuccessfulProjects

---

35

Build #EngineeringDesign schedules around a 50% confidence level as opposed to a 99%. If each task has 99% confidence, then the margin will be quite large and produce a very conservative schedule. #SuccessfulProjects

---

36

Build margin at the end of the
#EngineeringDesign program to improve
the confidence to higher than 50%.
#SuccessfulProjects

_____

_____

_____

37

A highly successful #EngineeringDesign
project's end schedule has a date range with
a confidence level. #SuccessfulProjects

_____

_____

_____

38

Get your stakeholders to come to an agreement on the #EngineeringDesign schedule, especially those who will have to contribute to meeting it. #SuccessfulProjects

_____

_____

_____

39

#EngineeringDesign stakeholders should include people from design, marketing, manufacturing, technical support, finance, sales, etc. #SuccessfulProjects

_____

_____

_____

40

Don't proceed without agreement on the #EngineeringDesign schedule. #SuccessfulProjects happen when agreements are in place.

_____

_____

_____

41

Ask for agreement on the #EngineeringDesign schedule, don't demand it. Make changes to get willing agreement from those who will work to meet the schedule. #SuccessfulProjects

_____

_____

_____

42

If stakeholders don't agree on the #EngineeringDesign schedule, don't assume they will agree later — they won't. There is either a flaw in the schedule, or you have not explained something well enough. #SuccessfulProjects

---

43

A good #EngineeringDesign schedule is challenging but doable. People enjoy a challenge, as long as it is not too difficult, and they get rewarded for success. #SuccessfulProjects

44

Beware of the person who objects to the #EngineeringDesign schedule but does not have a stake in it. Why is this person involved? #SuccessfulProjects

_____

_____

_____

45

"It shouldn't take this long" is an #EngineeringDesign trap. Trust your data over optimism to have #SuccessfulProjects.

_____

_____

_____

46

Identify #EngineeringDesign risks and plan to mitigate them. Allow for extra time in the schedule in proportion to the level of risk. #SuccessfulProjects

_____

_____

_____

47

To mitigate large #EngineeringDesign risks and have #SuccessfulProjects, do preliminary investigations, prototypes, pilots, or other testing.

_____

_____

_____

48

If you leave large #EngineeringDesign risks to be resolved during the project, allow time and budget for the predictable and inevitable "unknown" or "unexpected." #SuccessfulProjects

_____

_____

_____

49

#SuccessfulProjects with a detailed #EngineeringDesign plan make it easy to quickly revise the plan when changes happen during the project.

_____

_____

_____

50

If there is no #EngineeringDesign plan, the project will take far longer to complete than if there is a plan. #SuccessfulProjects happen when there's a plan.

_____

_____

_____

51

The budget for #EngineeringDesign projects should be planned as carefully as the schedule. #SuccessfulProjects

---

52

The budget for #EngineeringDesign projects is determined from the schedule by adding a cost for each person. #SuccessfulProjects

---

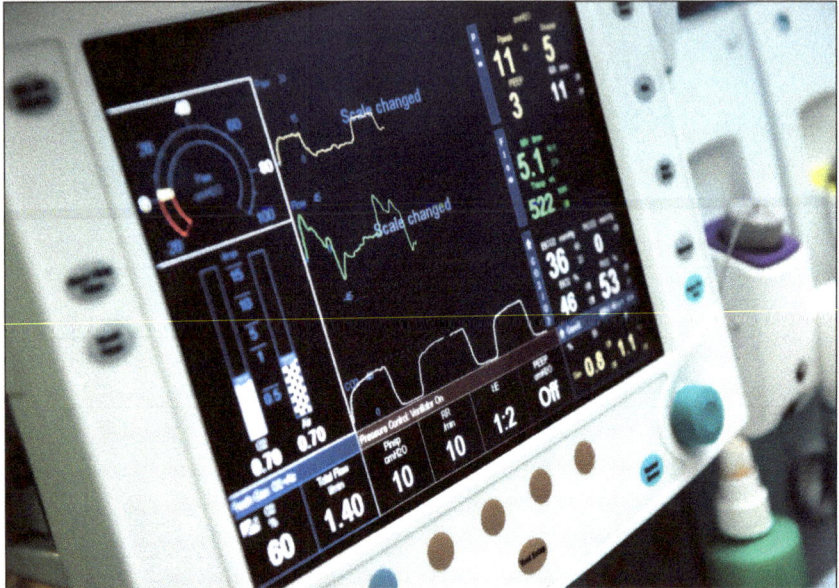

Make sure each medical device requirement is a single requirement, testable by a single test with a single outcome. #EngineeringDesign #SuccessfulProjects

Walt Maclay
http://aha.pub/EngineeringDesignProjects

Share the AHA messages from this book socially by going to
http://aha.pub/EngineeringDesignProjects

Section III

Medical Device Requirements

These requirements are important for medical device design projects, though most are a good idea for just about any project.

53

Medical device requirements serve as a good model for any good product design. If you're not doing a medical device, adjust the level of documentation to suit your particular needs. #EngineeringDesign #SuccessfulProjects

---

54

Successful medical #EngineeringDesign projects require documentation appropriate for the class of device (I, II, or III for FDA; A, B, or C for CE). Class III or C requires more documentation and gets more scrutiny. #SuccessfulProjects

---

55

Medical device development has a generally accepted flow: User requirements -> Engineering requirements -> Design -> Verification -> Validation. #EngineeringDesign #SuccessfulProjects

_____

_____

_____

56

Validation is usually done with animal testing or human testing (clinical trial). #EngineeringDesign #SuccessfulProjects

_____

_____

_____

57

Medical devices have several different types of requirements: 1) Product Requirements Document (PRD), 2) System Requirements (SysRS), and 3) Software Requirements Specification (SRS). #EngineeringDesign #SuccessfulProjects

_____

_____

_____

58

The Product Requirements Document (PRD) has the overall requirements and in simple medical devices, has the complete electrical and mechanical requirements. #EngineeringDesign #SuccessfulProjects

---

59

A complex medical device has System Requirements (SysRS), which are detailed requirements for electrical, mechanical, and industrial design and user experience. Simple devices usually don't have System Requirements. #EngineeringDesign #SuccessfulProjects

---

60

The Software Requirements Specification (SRS) has detailed software requirements. #EngineeringDesign #SuccessfulProjects

_____

_____

_____

61

List all the standards that apply to the medical device in the Product Requirements Document (PRD). #EngineeringDesign #SuccessfulProjects

_____

_____

_____

62

Provide an overview of the medical device and its uses that is separate from the requirements and will not be verified. #EngineeringDesign #SuccessfulProjects

_____

_____

_____

63

For medical devices, each #EngineeringDesign requirement will be tested during verification. Make sure each requirement can be tested with a measurable result. #SuccessfulProjects

_____

_____

_____

64

If you can't test a medical device requirement, it is either badly defined or should not be a requirement. Highly successful #EngineeringDesign projects have testable requirements. #SuccessfulProjects

_____

_____

_____

65

Make sure each medical device requirement
is a single requirement, testable by
a single test with a single outcome.
#EngineeringDesign #SuccessfulProjects

_____

_____

_____

66

Risk analysis will lead to new or changed
medical device requirements. Allow time in
the #EngineeringDesign schedule for this.
#SuccessfulProjects

_____

_____

_____

67

Medical device requirements sometimes change during the project. Always review the risk analysis, schedule, and budget if the requirements change. #EngineeringDesign #SuccessfulProjects

_____

_____

_____

68

Making changes in medical device requirements after verification testing is expensive. Ensure #SuccessfulProjects by making changes early or in the next version. #EngineeringDesign #SuccessfulProjects

_____

_____

_____

69

Making changes in medical device requirements after clinical testing is VERY expensive. Work hard to achieve successful clinical testing, and do pre-testing when possible to have a #SuccessfulProject. #EngineeringDesign

---

70

For more advice on medical device requirements, see: https://volersystems. com/product-design/medical-devices/ developing-product-requirements-medical-devices/. #EngineeringDesign #SuccessfulProjects

---

Have everyone, including junior people, report on their progress to ensure a highly successful #EngineeringDesign project. #SuccessfulProjects.

Walt Maclay
http://aha.pub/EngineeringDesignProjects

Share the AHA messages from this book socially by going to
http://aha.pub/EngineeringDesignProjects

Section IV

Essential Actions Required during Highly Successful Engineering Design Projects

Now that you have started off the design project right (see Section 2), this is what you need to do to continue to have a successful project.

71

Hold review meetings more frequently during crucial phases of the #EngineeringDesign project to achieve a highly successful project. #SuccessfulProjects

_____

_____

_____

72

In review meetings, discuss the progress, next steps, risks, schedule, and budget of the #EngineeringDesign project. #SuccessfulProjects

_____

_____

_____

73

During #EngineeringDesign meetings, focus on what's important. Don't waste everyone's time on something that can be better handled by an individual or a small group. #SuccessfulProjects

_____

_____

_____

74

Make sure important issues of the #EngineeringDesign project are covered completely in meetings. #SuccessfulProjects

_____

_____

_____

75

For a #SuccessfulProject, make
sure everyone contributes in the
#EngineeringDesign review meetings.

_____

_____

_____

76

Have everyone, including junior people,
report on their progress to ensure a highly
successful #EngineeringDesign project.
#SuccessfulProjects.

_____

_____

_____

77

Hold people accountable for the #EngineeringDesign schedule and budget to ensure #SuccessfulProjects. They agreed to it beforehand, didn't they?

_____

_____

_____

78

Work on the high-risk #EngineeringDesign tasks first. #SuccessfulProjects

---

79

Change is inevitable, don't fear it. Embrace change cautiously, but avoid making frequent changes to the #EngineeringDesign project. #SuccessfulProjects

---

80

Before making a change to an #EngineeringDesign project, list all the things that need to be done when that change is made, including the impact to the cost, schedule, and risk. #SuccessfulProjects

_____

_____

_____

81

To avoid last-minute changes, require all stakeholders to agree that a change is necessary and desirable in the #EngineeringDesign project. #SuccessfulProjects

_____

_____

_____

82

Don't micromanage. Highly successful #EngineeringDesign projects give people freedom to do things their way, but not so much freedom that failures can be large. #SuccessfulProjects

_____

_____

_____

83

Pay attention to what your people do well and not so well. Offer them specific, direct, and respectful feedback on how to improve on the #EngineeringDesign project. #SuccessfulProjects

_____

_____

_____

84

Expect your #EngineeringDesign staff to improve. Teach them and reward them, and they will improve. #SuccessfulProjects

_____

_____

_____

85

Are you rewarding your #EngineeringDesign team? Rewards make people feel valued. #SuccessfulProjects

_____

_____

_____

86

Reward your #EngineeringDesign people for taking necessary risks regardless of the outcome and for learning from mistakes. #SuccessfulProjects

_____

_____

_____

87

#SuccessfulProjects happen when you deliver specific, selective, and timely recognition in person, rather than by just rewarding with money. How do you reward your #EngineeringDesign staff?

_____

_____

_____

88

Learning how to improve is motivating and can lead to more successful #EngineeringDesign projects. #SuccessfulProjects

_____

_____

_____

89

Being on a successful #EngineeringDesign team is motivating. #SuccessfulProjects

_____

_____

_____

Realize that the definition of completion is different for every person. Make sure to come to common ground in the early stages of the #EngineeringDesign project, or there will be surprises later on. #SuccessfulProjects

Walt Maclay
http://aha.pub/EngineeringDesignProjects

Share the AHA messages from this book socially by going to
http://aha.pub/EngineeringDesignProjects

Section V

Warning Signs to Watch for during Highly Successful Design Projects

You are almost done, but don't forget that without successful manufacturing, the project is not successful. Engineering has a critical role in transferring the engineering design to manufacturing. Early in my career, it was pointed out to me that a design engineer does not create anything … except documentation that is used by others to create something. If the documentation is not good, the engineer's job was not well done.

90

Make sure the engineers understand the entire high-level product and not just their part of it. #EngineeringDesign #SuccessfulProjects

----

91

1 of 16 Warning Signs of #EngineeringDesign Problems: If people working on the project don't know what is expected of them. #SuccessfulProjects

----

92

2 of 16 Warning Signs of
#EngineeringDesign Problems:
If the schedule keeps changing.
#SuccessfulProjects

_____

_____

_____

93

3 of 16 Warning Signs of #EngineeringDesign Problems: If the requirements keep changing. #SuccessfulProjects

_____

_____

_____

94

4 of 16 Warning Signs of #EngineeringDesign Problems: If the budget keeps changing. #SuccessfulProjects

_____

_____

_____

95

5 of 16 Warning Signs of #EngineeringDesign Problems: If the requirements are never allowed to change. You must adapt to changes, such as competitors or technical discoveries. #SuccessfulProjects

_____

_____

_____

96

6 of 16 Warning Signs of #EngineeringDesign Problems: If manufacturing isn't involved in the design. #SuccessfulProjects

_____

_____

_____

97

7 of 16 Warning Signs of
#EngineeringDesign Problems: If
marketing and sales aren't involved in the
design. #SuccessfulProjects

_____

_____

_____

98

8 of 16 Warning Signs of #EngineeringDesign Problems: If technical support is not involved in the design. #SuccessfulProjects

_____

_____

_____

99

9 of 16 Warning Signs of #EngineeringDesign Problems: If engineers on the project have different understandings of what the product is. #SuccessfulProjects

_____

_____

_____

100

10 of 16 Warning Signs of #EngineeringDesign Problems: If engineers on the project don't understand who the customers and users are and what they want. #SuccessfulProjects

_____

_____

_____

101

11 of 16 Warning Signs of #EngineeringDesign Problems: If design engineers are not involved in the transfer to manufacturing. #SuccessfulProjects

_____

_____

_____

102

12 of 16 Warning Signs of #EngineeringDesign Problems: If review meetings usually have arguments. #SuccessfulProjects

_____

_____

_____

103

13 of 16 Warning Signs of #EngineeringDesign Problems: If there are few or no review meetings. #SuccessfulProjects

_____

_____

_____

104

14 of 16 Warning Signs of #EngineeringDesign Problems: If only design engineers attend review meetings. #SuccessfulProjects

_____

_____

_____

105

15 of 16 Warning Signs of #EngineeringDesign Problems: If management doesn't know the key risks in the project. #SuccessfulProjects

_____

_____

_____

106

16 of 16 Warning Signs of #EngineeringDesign Problems: If management doesn't review the actual versus planned schedule and budget regularly. #SuccessfulProjects

_____

_____

_____

107

Realize that the definition of completion is different for every person. Make sure to come to common ground in the early stages of the #EngineeringDesign project, or there will be surprises later on. #SuccessfulProjects

_____

_____

_____

Any #EngineeringDesign issue found needs to be included in an updated manufacturing test. Check that the manufacturing test will identify devices with each issue that is found. #SuccessfulProjects

Walt Maclay
http://aha.pub/EngineeringDesignProjects

Share the AHA messages from this book socially by going to
http://aha.pub/EngineeringDesignProjects

Section VI

Ensuring a Successful Transfer to Manufacturing

It's easy to have problems that go unnoticed, especially when you are still learning to have highly successful design projects. These are things to watch out for.

108

Highly successful #EngineeringDesign projects have complete requirements to ensure a successful transfer to manufacturing. #SuccessfulProjects

_____

_____

_____

109

In a highly successful #EngineeringDesign project, it is easy to buy the parts, build it, test it, ship it, and support it in the volumes required and to the quality specifications. #SuccessfulProjects

_____

_____

_____

110

Manufacturing needs tightly controlled #EngineeringDesign documentation that is uniform and easy to follow and repeat without a lot of training. #SuccessfulProjects

---

111

1 of 2 disconnects between engineering and manufacturing that need solving to have successful #EngineeringDesign Projects: Engineers usually don't understand why the transfer to manufacturing should take so much of their time. #SuccessfulProjects

---

112

2 of 2 disconnects between engineering and manufacturing that need solving to have successful #EngineeringDesign Projects: Manufacturing people usually don't understand why engineers are so sloppy and hard to deal with. #SuccessfulProjects

_____

_____

_____

113

Manufacturing needs these standard #EngineeringDesign documents: Bill of Materials, Gerber files, executable software files, dimensioned drawings, assembly instructions, and test instructions to ensure #SuccessfulProjects.

_____

_____

_____

114

Manufacturing needs manufacturable #EngineeringDesigns in addition to the standard documents. #SuccessfulProjects

_____

_____

_____

115

Manufacturable #EngineeringDesigns start with good requirements. #SuccessfulProjects

_____

_____

_____

116

Manufacturable #EngineeringDesigns
have manufacturing in mind from the start.
#SuccessfulProjects

_____

_____

_____

117

Manufacturable #EngineeringDesigns are easy to assemble. #SuccessfulProjects

_____

_____

_____

118

Manufacturable #EngineeringDesigns have careful part selections that are affordable and procurable. #SuccessfulProjects

_____

_____

_____

119

Manufacturable #EngineeringDesigns require careful design to be able to meet specifications, despite part variations, to ensure #SuccessfulProjects.

120

Manufacturable #EngineeringDesigns are easy to test. #SuccessfulProjects

---

121

Manufacturable #EngineeringDesigns have test points in the printed circuit board (PCB) layout to facilitate testing. #SuccessfulProjects

_____

_____

_____

122

Manufacturable #EngineeringDesigns consider testing during the design. #SuccessfulProjects

_____

_____

_____

123

Manufacturable #EngineeringDesigns get input from manufacturing, so they are suited to the manufacturing process of the manufacturer. #SuccessfulProjects

_____

_____

_____

124

Manufacturable #EngineeringDesigns allow for parts to be easily removed from molds by having draft. #SuccessfulProjects

_____

_____

_____

125

Manufacturable #EngineeringDesigns do not get last-minute changes in the requirements. #SuccessfulProjects

_____

_____

_____

126

Expect that the transfer to manufacturing will have a few issues. Plan this into your schedule. Remember that building #EngineeringDesign prototypes is quite different from building goods ready to sell. #SuccessfulProjects

_____

_____

127

A 10x increase in manufacturing volume brings out a new set of #EngineeringDesign issues to be addressed. Plan on time to find and address these issues. #SuccessfulProjects

_____

_____

_____

128

When product issues are found, there needs to be revised drawings, work instructions, etc. There needs to be an outcome for any issue found, or it is being ignored. #SuccessfulProjects

_____

_____

_____

129

Any #EngineeringDesign issue found needs to be included in an updated manufacturing test. Check that the manufacturing test will identify devices with each issue that is found. #SuccessfulProjects

_____

_____

_____

130

Use bad parts in the manufacturing test to ensure that the test procedure finds all issues. #SuccessfulProjects

_____

_____

_____

131

Part availability needs to be monitored during the life of the product. End-of-life parts can be a large #EngineeringDesign issue. Stay on top of this to mitigate a potential line-down situation. #SuccessfulProjects

_____

_____

_____

If you don't do anything different with your #EngineeringDesign projects, you won't get different results. #SuccessfulProjects

Walt Maclay
http://aha.pub/EngineeringDesignProjects

Share the AHA messages from this book socially by going to
http://aha.pub/EngineeringDesignProjects

Section VII

Integrating Learning to Get Better with Every Project

No matter what level of success you just had, learn how to improve. With these steps, a good team can become great.

132

Get better on every #EngineeringDesign project by improving processes, systems, tools, and practices based on what you learned. #SuccessfulProjects

_____

_____

_____

133

Focus on what is required beyond the #EngineeringDesign shipment phase. #SuccessfulProjects

_____

_____

_____

134

Highly successful #EngineeringDesign projects design for supportability. #SuccessfulProjects

_____

_____

_____

135

Highly successful #EngineeringDesign projects design for repairability. #SuccessfulProjects

_____

_____

_____

136

Review the #EngineeringDesign project with your customer(s), and look for constructive feedback. #SuccessfulProjects

_____

_____

_____

137

Highly successful #EngineeringDesign projects encourage criticism by suggesting possible things you could have done better. This also encourages praise where it is due. #SuccessfulProjects

_____

_____

_____

138

Review the #EngineeringDesign project with the team. Include the extended team and all the parties involved in the project. Present feedback from customers. #SuccessfulProjects

_____

_____

_____

139

Solicit constructive criticism from the team. Make sure all members are heard. #EngineeringDesign #SuccessfulProjects

_____

_____

_____

140

Update #EngineeringDesign procedures and practices from what is learned by reviewing the project. #SuccessfulProjects

_____

_____

_____

About the Author

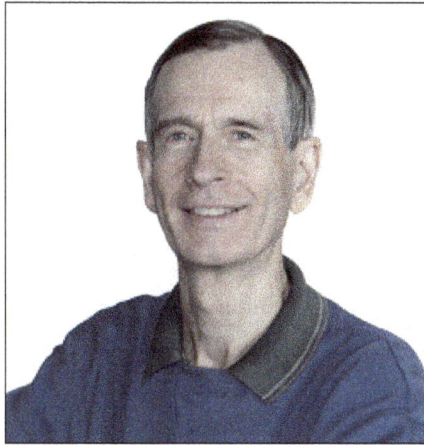

Walt Maclay is the president and founder of Voler Systems in 1979, now one of the top electronic design firms in Silicon Valley, and is committed to delivering quality electronic products that are easy to manufacture on time and on budget. Voler Systems provides design, development, risk assessment, and verification of new devices for medical, consumer, and industrial applications. Voler is particularly experienced in designing wearable and IoT devices, using its skill with sensors and wireless technology.

Mr. Maclay has been active in several consultant organizations and is a senior life member of the Institute of Electrical and Electronic Engineers (IEEE). He is a reviewer for NSF SBIR grants and has mentored dozens of startup companies. Voler Systems is a member of a technology consortium, the Product Realization Group, which provides all the services to design and introduce new hardware products. Mr. Maclay holds a BSEE degree in electrical engineering from Syracuse University.

Contributors:

John Dring, VP of Engineering at Voler Systems, contributed his great knowledge and experience.

Kimberly Wiefling, founder and president of Wiefling Consulting, edited and contributed a great deal.

AHAthat®

THiNKaha has created AHAthat for you to share content from this book.

- ⮑ Share each AHA message socially:
  http://aha.pub/EngineeringDesignProjects
- ⮑ Share additional content: https://AHAthat.com
- ⮑ Info on authoring: https://AHAthat.com/Author

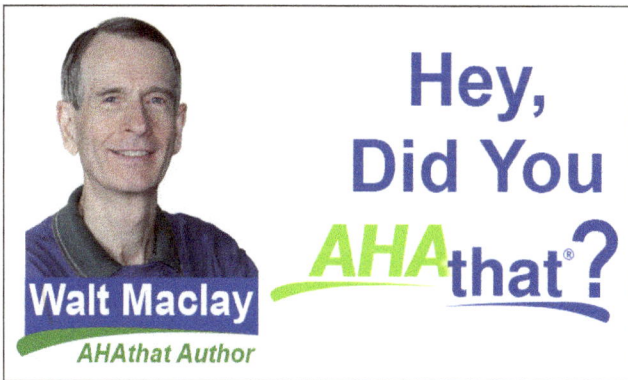

Hey, Did You **AHA**that®?

Walt Maclay
AHAthat Author